傑克的雲端探險

Jack's Cloud Adventure

從前，有個叫傑克的小男孩，他和媽媽還有一頭乳牛住在一起，過得很辛苦。

Once upon a time, there was a boy named Jack, who lived with his mother and a cow. They were very poor and lived a hard life.

3

有一天，乳牛再也擠不出牛奶了，
媽媽就叫傑克到鎮上把乳牛賣掉。

One day, their cow stopped producing milk,
so Jack's mother sent Jack off to the town to sell the cow.

傑克走著走著，來到了一條大河旁，
可是這條大河上沒有橋，傑克該怎麼渡河呢？你可
以幫傑克畫一座橋嗎？

Jack walked and walked, and came to a big river,
but there was no bridge to cross this big river.
What should Jack do? Can you help Jack draw a bridge?

謝謝你，傑克順利的渡河，
他可以繼續往鎮上出發了。

Thank you, Jack passed the river
successfully.
He can continue to set off for town.

途中，傑克遇到一個男人。
Along the way, Jack met a man.

「我手上有幾顆神奇的魔豆，用來跟你交換乳牛好不好？」

"I have some magic beans that I would like to trade for your cow?"

「魔豆？那一定很神奇！」
傑克開心的想著。

"Magic beans? That sounds awesome!" thought Jack.

「好啊！
我跟你交換。」

"Yes! I will trade with you."

傑克開心的帶著魔豆回家去了。

Jack happily went home with the magic beans.

回到家後，媽媽發現傑克居然用乳牛交換了幾顆豆子，沒有賺到錢，生氣的將豆子丟出窗外。

After Jack got home, his mother found that Jack had traded the cow for some beans without making any money. So she furiously threw the beans out the window.

隔天，
被丟到窗外的魔豆，
圓圓的殼破了，
冒出小小的芽。

The next day, the bean that was thrown out of
the window had a small bud break through it's
round surface.

第二
豆子發
細細的根紧
準備

On the second d
Its thin roots tigh
ready to gr

天，
⋯了，
⋯抓住土壤，
⋯大。

⋯e bean sprouted.
⋯ng to the soil,
⋯up and out.

第三天，
豆子細細白白的莖
從土壤冒出來，
並長出小小的葉子。

On the third day, a thin white stem emerged
from the soil and small leaves began growing.

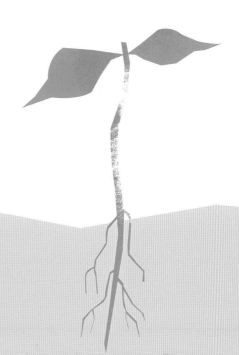

傑克開心極了，
每天細心的照顧豆子，為它澆水。

Jack was very happy, and he carefully took care of the bean sprouts and
watered them every day.

豆子一天一天長高，越長越高、越長越高⋯⋯直到有一天，魔豆長到天上去啦！好奇的傑克順著豆子的莖往上爬，爬到了雲端。

Day after day, the beanstalk grew taller...and taller...and taller...until one day, the beanstalk had grown into the sky! Jack climbed up the beanstalk up into the clouds.

15

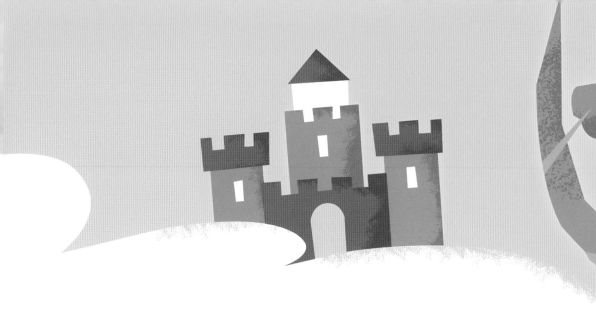

想不到雲端上有一大片田地，傑克摸摸口袋，發現還有幾顆豆子在口袋裡，決定把它們灑在田地裡，並且天天到雲端上替它們澆水。

Unexpectedly, there was a large field in the clouds. Jack felt his pockets and found that there were still a few beans left in his pocket. He decided to sprinkle them in the field and watered them up in the clouds every day.

雲端上種植出來的植物，比地上種的更好吃，最後豆子們長成了茂密的作物。

The crops gathered from the farmland upon the clouds tasted far better than the crops on earth. The magic beans grew into a dense field of crops.

看一看
Look

摸一摸
Touch

小朋友，你喜歡吃蔬菜水果嗎？
分享你喜歡的蔬果，說說它們的形狀、
顏色和味道吧！

Friends, do you like vegetables and fruits?
Share your favorite fruits and vegetables, and describe their
shapes, colors and tastes!

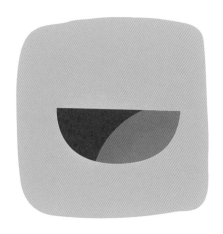

聞一聞
Smell

嘗一嘗
Taste

傑克賣掉這些農作物，賺了好多錢，
從此和媽媽過著幸福快樂的日子。

Jack sold these crops and made a lot of money. From then on,
he lived happily ever after with his mother.

生字練習
Vocabulary

1. 豆ㄉㄡˋ 子ㄗ˙
   beans

2. 發ㄈㄚ 芽ㄧㄚˊ
   germinate

3. 根ㄍㄣ
   root

4.

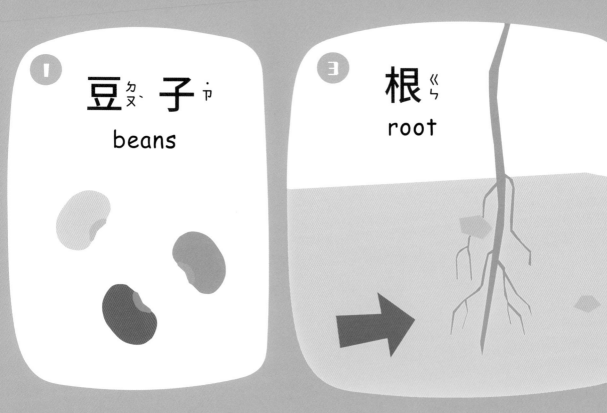

5

莖 ㄐㄧㄥ
stem

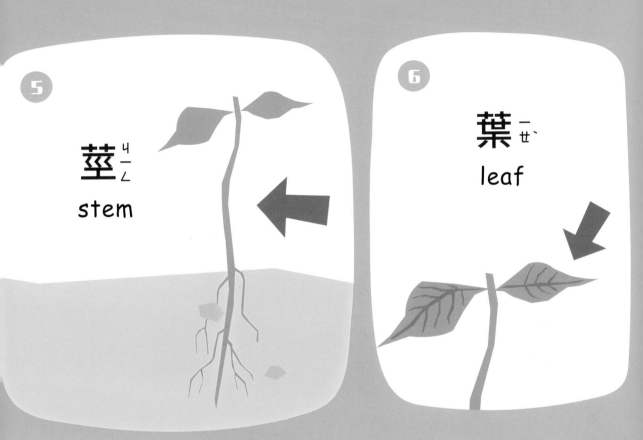

6

葉 ㄧㄝˋ
leaf

土 ㄊㄨˇ 壤 ㄖㄤˇ
soil

7

澆 ㄐㄧㄠ 水 ㄕㄨㄟˇ
watering

神農氏

Shennong

很久很久以前，人們靠著捕食野獸為生，可是人太多，野獸漸漸的不夠人們吃了。還好，這時出現一個叫神農氏的人，他教導人們製作農具耕田，並教人們如何種植作物，這樣一來人們就不會餓肚子了。除此之外，神農氏還親自嘗了各種藥草，把藥草的功效記錄下來，讓人們可以依此治病。

A long time ago, people lived on predating beasts. However, there were too many people, and gradually the beasts were not enough for people to eat. Fortunately, there was a person named Shennong who taught people how to make farm tools and grow crops, so that people would not suffer from hunger because of short of beasts. Shennong also personally tasted various herbs and recorded the medical values so that people could use herbs to treat the disease.

小朋友，看完故事後，
跟著傑克一起唱唱歌吧！
掃 QRcode 跟著旋律一起唱吧！

Children, after reading the story, let's sing along with Jack! Scan the QRcode and sing along with the melody!

傑克的小魔豆
Jack's Little Magic Bean

傑克有顆小豆子他種到土裡去
　　　　鬆鬆土壤澆澆水
　　　　慢慢長大啦

　　　　發芽叮叮叮
　　　　豆莖咚咚咚
　　　　葉子唰唰唰
　　　　啦啦啦啦啦

傑克有顆小豆子它長到天上去
　　　　走走雲上撒撒豆
　　　　慢慢長大啦

　　　　發芽叮叮叮
　　　　豆莖咚咚咚
　　　　葉子唰唰唰
　　　　啦啦啦啦啦

27

ACTIVITY TIME!!

葉子的拓印
Leaf Print

我們來當小小觀察家，找一找落葉，仔細清理並觀察後，拿來做美麗的拓印吧！

Let's be little observers, find a fallen leaf, carefully clean and observe it, then let's use it to make a beautiful print!

動手做 1：葉子的拓印
Handicraft 1: Leaf Print

材料：葉子、紙、印泥（水彩顏料）、印章
Material: leaf, paper, ink(watercolor paints),stamp

1

選擇喜歡的印泥顏色。

Choose your favorite ink color.

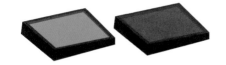

2

把清洗好的葉子沾上印泥。

Dip the clean leaf in the ink.

3

再把葉子印在紙上。

Place the leaf on the paper. Press done.

4 蓋上印章做裝飾，美麗的葉子拓印就完成啦！

Use it as a stamp to make a decoration, the beautiful leaf prints are done!

可拓印於家長引導指南 p14.
Kids can play it with parents guide P.14

29

不同的豆豆
Different Kinds of Beans

豆豆的種類真多，它們都長得不一樣。我要拿著放大鏡好好的觀察它們，跟大家分享我所看到的豆豆。

There are so many types of beans, they all look so different. I am going to use a magnifying glass to examine them carefully and share with everyone what I see.

材料：綠豆、黃豆、紅豆、黑豆
Material: Mung beans, soy beans, red beans, black beans.

動手做2：豆豆沙鈴
Handicraft 2: Bean Maracas

材料：兩個養樂多瓶、透明膠帶、色紙、亮片
Material: Two Yakult bottles, tape, colored paper, sequins

內容物：綠豆、黃豆或是米
Contents: mung beans, soy beans or rice.

1

將綠豆放進養樂多瓶。

Put mung beans in the bottles.

2

把瓶口用膠帶封起來。

Seal the opening of the bottle with tape.

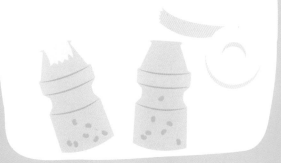

3

把兩個養樂多瓶用膠帶貼起來。

Put the ends of the bottles together and tape them together.

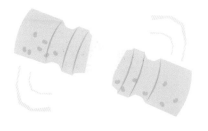

4

用色紙、亮片裝飾你的豆豆沙鈴吧！

Use the colored paper and sequins to decorate your maracas.

做好沙鈴後，請翻回 27 頁，一起搖擺沙鈴
When you're done making your maracas, go to page 27 and let's shake our maracas together!

ACTIVITY TIME!!

動手做 3：種種小魔豆
Handicraft 3：Planting My Little Magic Beans

材料：花盆、土壤、種子、水壺
Material: flower pot, soil, seed, watering can

1 先倒入半盆土。
Fill the pot half way with soil.

2 將種子撒在花盆內。
Sprinkle the seeds in the pot.

3 覆蓋上土壤。
Cover with soil.

4 澆澆水。
Watering.

種下魔豆，記得要讓魔豆曬曬太陽並天天澆水喔。

Plant the magic beans, remember to let the magic beans soak up some sun and
get some water every day.

★可觀察魔豆的成長過程，紀錄在家長手冊 p.12。

動手做4：雲端上的城堡
Handicraft 4: Castle on the cloud

材料：花盆、衛生筷、彩色筆、膠帶、附件紙張
Materials: flower pot, disposable chopsticks, colored pens, page from the back of the book.

1 將附件的城堡塗上喜歡的顏色。
Paint the castle (from the back of the book) in your favorite color.

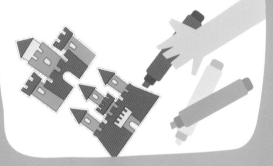

2 用剪刀沿著虛線將圖片剪下。
Cut the picture along the dotted line with scissors.

3 將剪下的城堡貼在衛生筷上。
Tape the picture to the chopstick.

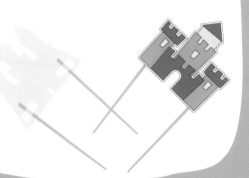

4 將竹筷插進花盆裡，雲端城堡就完成了！
Insert the chopstick into the flowerpot, and your Castle in the clouds is complete!

33

傑克要去市集賣蔬果，但他卻迷路了，你能幫忙他找到市集嗎？

Jack is going to the market to sell fruits and vegetables, but he is lost. Can you help him find the market?

會喝水的蔬菜
Vegetables can drink

小朋友，你知道神奇的蔬菜會喝水嗎？

我們一起來做個實驗吧！

Children, did you know that magic vegetables drink water?
Let's do an experiment together!

動手做ㄅ: 會喝水的蔬菜
Handicraft 5: Vegetables can drink

材料：芹菜兩根、兩杯水、食用色素
Ingredients: Two stalks of celery, two glasses of water, food coloring

1
先把一根芹菜的葉子拔光。
First take all the leaves off of the celery.

2
將食用色素滴進水裡。
Drip some food coloring into the water.

3
把兩根芹菜分別放進杯子裡。
Put a stalk of celery in each cup.

4
觀察葉子的顏色、水量，並把觀察結果記錄下來。
Note the color of the stalks, the amount of water in the cup, and record your observations.

小朋友，你發現了什麼呢？想知道答案的話，快掃描 QRcode 吧！

What did you find out? If you want to know the answer, scan the QRcode.

在家玩 STEAM+C

傑克的雲端探險 Jack's Cloud Adventure

文‧圖／ 你好泡泡 nhbubble

你好泡泡編輯團隊／ Ni Hao Chinese LLC

副主編／胡琇雅 美術編輯／ 鍾菱、沈昭憲

董事長／趙政岷 第五編輯部總監／梁芳春

出版者／時報文化出版企業股份有限公司

　　　　10803 台北市和平西路三段 240 號七樓

發行專線／（02）2306-6842

讀者服務專線／0800-231-705、（02）2304-7103

讀者服務傳真／（02）2304-6858

郵撥／ 1934-4724 時報文化出版公司

信箱／10899 台北華江橋郵局第99信箱

統一編號／ 01405937

時報悅讀網／ www.readingtimes.com.tw

法律顧問／理律法律事務所 陳長文律師、李念祖律師

Printed in Taiwan

初版一刷／ 2020 年 02 月 14 日

時報文化出版公司成立於一九七五年，並於一九九九年股票上櫃公開發行，
於二〇〇八年脫離中時集團非屬旺中，以「尊重智慧與創意的文化事業」為信念。

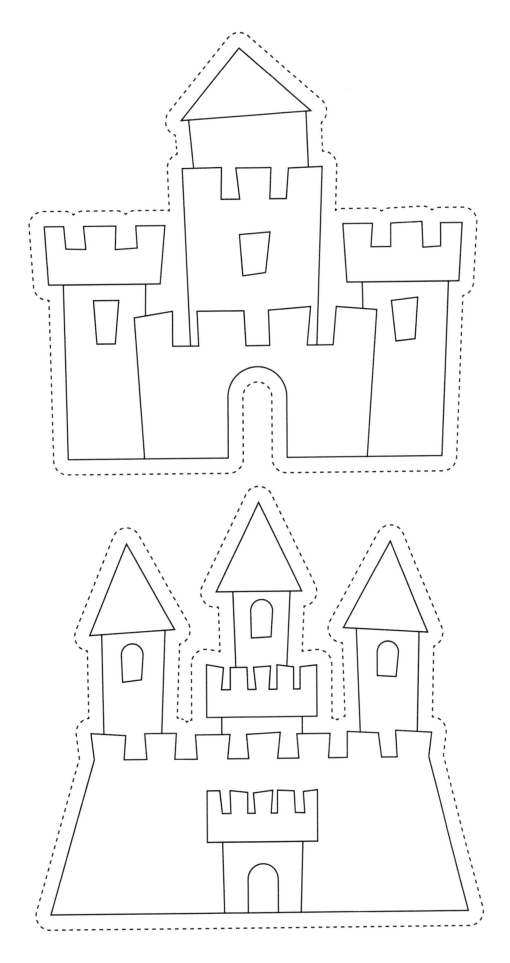